T0224985

SpringerBriefs in Architectural Design and Technology

Series Editor

Thomas Schröpfer, Architecture and Sustainable Design, Singapore University of Technology and Design, Singapore, Singapore

Indexed by SCOPUS

Understanding the complex relationship between design and technology is increasingly critical to the field of Architecture. The *Springer Briefs in Architectural Design and Technology* series provides accessible and comprehensive guides for all aspects of current architectural design relating to advances in technology including material science, material technology, structure and form, environmental strategies, building performance and energy, computer simulation and modeling, digital fabrication, and advanced building processes. The series features leading international experts from academia and practice who provide in-depth knowledge on all aspects of integrating architectural design with technical and environmental building solutions towards the challenges of a better world. Provocative and inspirational, each volume in the Series aims to stimulate theoretical and creative advances and question the outcome of technical innovations as well as the far-reaching social, cultural, and environmental challenges that present themselves to architectural design today. Each brief asks why things are as they are, traces the latest trends and provides penetrating, insightful and in-depth views of current topics of architectural design. *Springer Briefs in Architectural Design and Technology* provides must-have, cutting-edge content that becomes an essential reference for academics, practitioners, and students of Architecture worldwide.

Joerg Baumeister · Despina Linaraki

Cities+1m

Urban Development Solutions for Sea-Level Rise

 Springer

Joerg Baumeister
SeaCities
Griffith University
Gold Coast Campus, QLD, Australia

Despina Linaraki
SeaCities
Griffith University
Gold Coast Campus, QLD, Australia

ISSN 2199-580X   ISSN 2199-5818   (electronic)
SpringerBriefs in Architectural Design and Technology
ISBN 978-981-19-1375-4   ISBN 978-981-19-1376-1   (eBook)
https://doi.org/10.1007/978-981-19-1376-1

This Springer imprint is published by the registered company Springer Nature Singapore Pte Ltd.
The registered company address is: 152 Beach Road, #21-01/04 Gateway East, Singapore 189721, Singapore

Contents

Introduction

No rational individual is questioning global warming anymore. Coastal cities are already impacted due to effects like sea-level rise (SLR) and intensification of flooding events. Thereby, a precise prediction of the future impacts' extent is currently still impossible. Ice sheet melting processes due to increasing temperatures, for example, are still poorly understood. Looking at history with a comparable amount of greenhouse gases tells us that the sea can rise 15-25m in the next hundreds of years (Jones, 2017).

Humankind's potential change of behaviour won't have any recovery effects in the next decades (Samset et al., 2020). Therefore predictions of up to 1.40m or more SLR until 2100 (Oppenheimer et al., 2019) should be taken very seriously, and estimations are still developing upwards. Due to this insecurity, this book is not focusing on a particular year but on a specific height of +1m (1m above current highest astronomical tide) which derives from SLR and other potential flooding events that can occur because of stormwater run-off, rising groundwater table or river overflow.

+1m will change the coastal environment and impact elements of coastal cities. Infrastructure, buildings, industry, and communities will have to be protected, or cities have to retreat. Both concepts, protection, and retreat, are often unfavourable. The former needs elements like dykes which are costly and temporarily, the latter abandons existing urban assets and communities. Two more concepts are perhaps sometimes more promising: Implemented examples demonstrate that coastal cities can also advance onto the water or, like the HafenCity in Hamburg, they can accommodate. The intention of this publication is to consider all concepts.

Example of HafenCity Hamburg demonstrating the accommodation concept and protecting the city from 8m storm floods.

The guideline

Protect, Retreat, Advance, Accommodate: Which concepts should be applied to transform coastal cities to +1m and who will make the decisions?

There are no simple answers to these questions, neither regarding the only right concept nor the one decider. Therefore we developed a guideline that considers flexibility of concepts and decision making with the aim of creating multiple scenarios.

This guideline should be tested in various contexts of coastal cities, promoting sustainable planning guidance like inclusive and participative processes; the implementation of sustainable urban design scenarios; the promotion of knowledge exchange and cooperation between government sectors, the business community, and private citizens; the stimulation of education, information, and research in urban design for sustainability and sustainable urban development (Chrysoulakis et al., 2015).

This guideline could have the potential to stimulate also the required integration of government and community into research and education in fields like urban design and planning, engineering, infrastructure, ecology, economy, and others.

Risk-Management is thereby not the primary objective. On the opposite, we want to demonstrate that the transformation of coastal cities can create more opportunities than threats. Alternatively to multi-Million Dollar investments for coastal protection, this guideline will consider more sustainable urban, green, and blue development options.

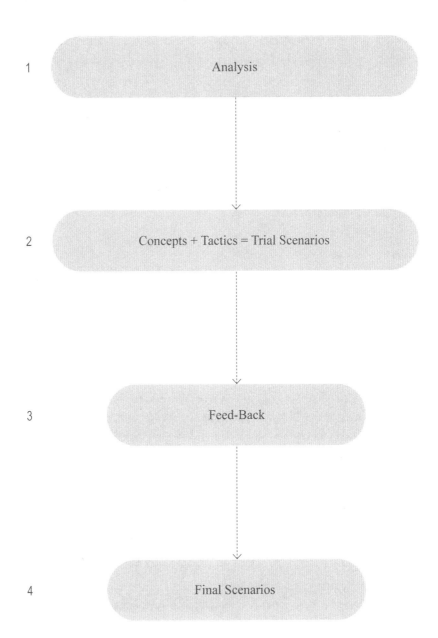

The methodology

The methodology follows a particular procedure based on observational and data-driven analysis, the creation, critical review, and improvements of experimental concepts and their simulations.

To achieve the best opportunities, this methodology considers the specific conditions of the examined site, including its current and future water-hazards, as well as the parallel development of state of the art scenarios on-site, the feed-back from stakeholders, and the final visualisation and description of the chosen "finalists".

1. This results in a logical step-by-step approach that starts with the site analysis by focusing on the collection of relevant data regarding the site and potential flooding areas which allows afterwards an evaluation of risks and opportunities.

2. It continues with the creation and visualisation of trial development scenarios by introducing known concepts and tactics for the site. Concepts relate in this case to concepts of the Urban Adaptation Models for SLR Adaptation (Baumeister, 2020). Tactics reflect a single activity or technique. Scenarios consider large, overarching plans in which tactics are coordinated (Cambridge dictionary, 2020) This creates systematically four water-adaptive urban development options and provides the site at the same time with additional water-adaptive assets.

3. Afterwards the second step is evaluated by a stakeholder feed-back. Participants will get thereby the chance to comment on existing ideas and to suggest additional ones on different scales.

4. The final step visualises and describes the best development scenarios. Each of them will highlight a different water-adaptive concept with relevant urban design features.

An additional expected outcome of this methodology is (in contrast to a normal Masterplan) its flexibility in terms of decision making and adaptability for future demands of the guideline.

The success of these advantages will be assessed at the end of the test run which will follow after the upcoming detailed description of the methodology.

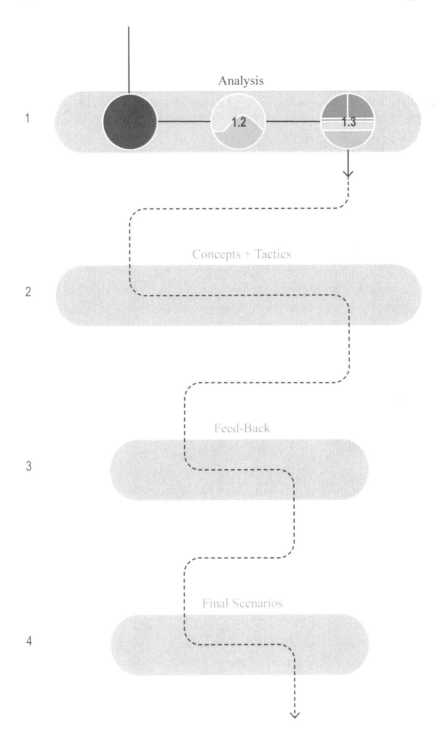

Steps 1.1 - 1.3

Following objectives are included in the Analysis:

1.1. Site Analysis

1.1.1. Site location: Select the site boundaries and collect relevant data (GIS, Google Earth, etc)

1.1.2. Historical development: Study the history of the site (plans, photos) including eventual water interactions like flooding

1.1.3. Site visit: Visit the site and take site photos that present potential flooding points and areas

1.1.4. Buildings, vegetation, elevation: Check buildings, vegetation types, and heights and indicate contour lines

1.2. Analysis of potential flood risks

1.2.1. Impact of SLR: Understand the impact of SLR / storm-water (past, present, and future)

1.2.2. Future flood levels: Simulate future water levels (e.g. +0.2m, +0.5m, +0.8m, and +1m)

1.2.3. Risk map: Create a risk map that indicates risk levels of buildings, infrastructure, and vegetation

1.3. Analysis of risks and opportunities of the site

1.3.1. Subdivision of site: Divide the site into tangible parts with different functions and flooding levels

1.3.2. Risks and opportunities: Analyse the Urban Elements (Baumeister and Ottmann, 2014) that will be described in the test run and will compile the risks and opportunities

1.3.3. Conclusion of step 1: Review critically step 1

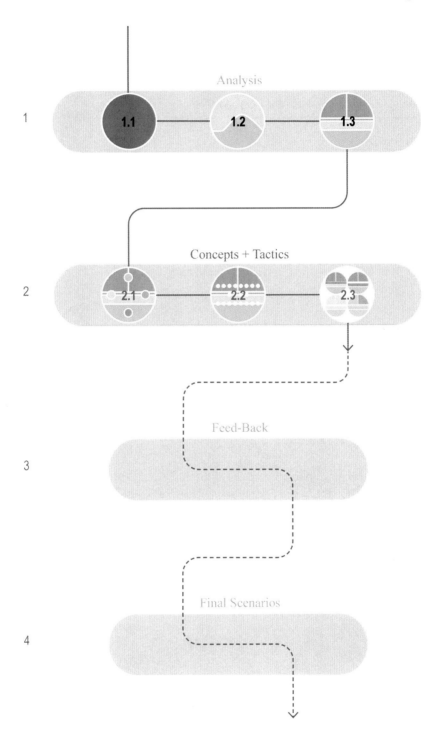

Steps 2.1 - 2.3

Step two is based on the introduction of concepts and tactics (Chrysoulakis et al., 2015) which will be described comprehensively during the test run:

2.1. Concepts (Urban Development Scale)
2.1.1. Adaptation Methods: Introduce and test adaptation methods on site
2.1.2. Combinations: Mix adaptation methods to increase the benefits
and evaluate corresponding opportunities

2.2. Tactics (Urban Design Scale)
2.2.1. Tactics' logos: Introduction twenty tactics which can be relevant
for the site
2.2.2. Tactics' explanations: Describe the different tactics

2.3. Concepts + Tactics = Scenarios
2.3.1. Opportunity A: Choose, visualise and describe the most logical mix
of concepts and tactics for opportunity 1
2.3.2. Opportunity B: re 2.3.1.
2.3.3. Opportunity C: re 2.3.1.
2.3.4. Opportunity D: re 2.3.1.
2.3.5. Conclusion of step 2

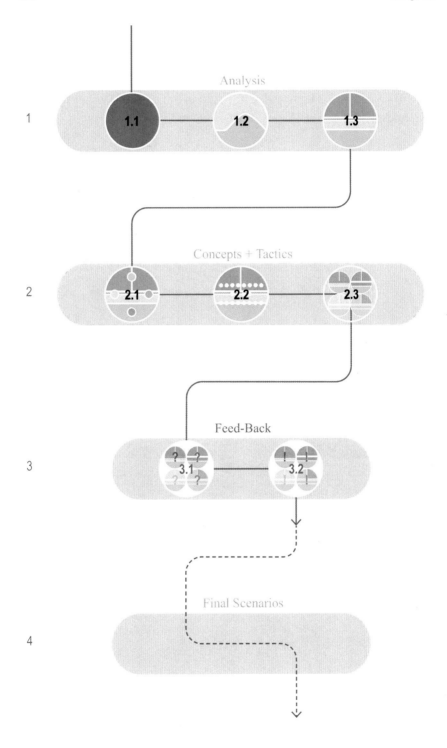

Steps 3.1 - 3.2

Step three asks for feed-back from stakeholders and the community to develop the proposed trial scenarios further:

3.1. Survey
   3.1.1. Survey of Adaptation methods: Ask for the advantages and disadvantages of each development concept for the specific site
   3.1.2. Survey of Tactics: Explain and ask for preferred development tactics
   3.1.3. Survey of Trial Scenarios: Ask for the pros and cons of each presented development scenario and collect additional personal ideas from each participant

3.2. Results
   3.2.1. Results of Adaptation methods: Collect the results of the invited stakeholder group
   3.2.2. Results of Tactics: Collect the results of invited stakeholders
   3.2.3. Results of Trial Scenarios: Collect the results of the invited stakeholders and check eventual discrepancies of results between 3.2.1. and 3.2.3.
   3.2.4. Conclusion of step 3: Evaluate the outcome of the survey regarding preferences and gaps of the proposed options to create in the next step prioritised opportunities for the site

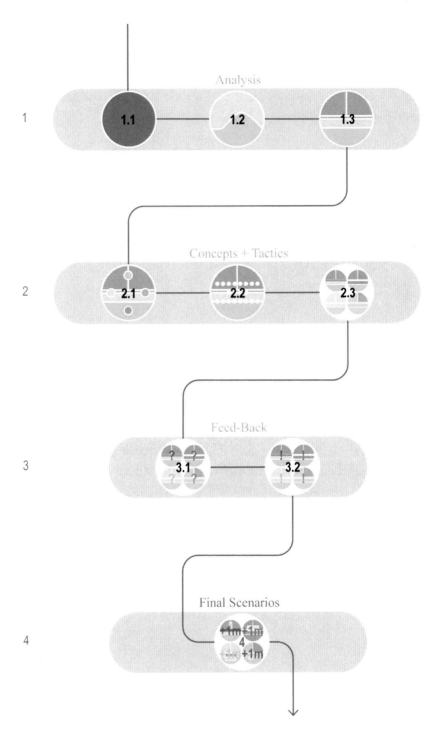

Step 4.

This final step results in urban development scenarios for sea-level rise:

4. Cities+1m
 4.1. Existing conditions
 4.2. Scenario **A** "Minimal Change": Visualise the improved scenario as a site plan and in 3D
 4.3. Scenario **B** "Maximal Yield": Visualise the improved scenario as a site plan and in 3D
 4.4. Scenario **C** "Maximal Nature": Visualise the improved scenario as a site plan and in 3D
 4.5. Scenario **D** "Maximal Water": Visualise the improved scenario as a site plan and in 3D

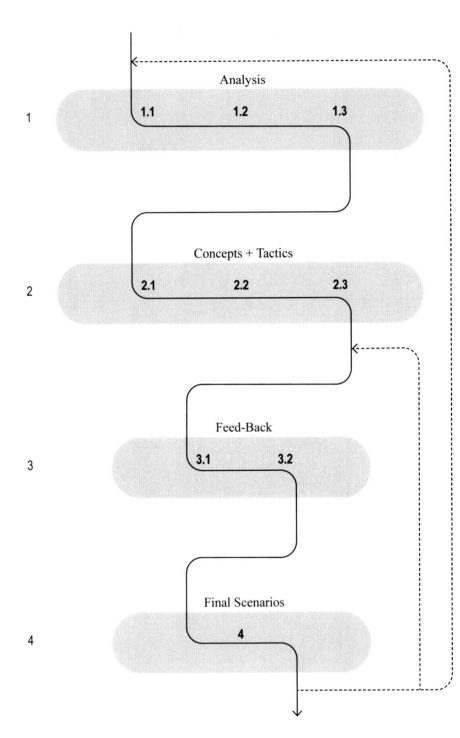

Conclusion

This guideline has been created to generate scenarios that transform coastal cities for sea-level rise. The step-by-step approach seems to be logical but has to be tested as a test run on a specific site.

The analysis of relevant data for a specific site and the evaluation of risks and opportunities follow the normal procedure (step 1). Due to various interpretations in literature, existing concepts for sea-level rise will be redefined and reinforced by twenty tactics which will create four scenarios (step 2).

The four proposed scenarios will be presented and discussed with stakeholders and the community (step 3) before the feed-back will be compiled in final scenarios (step 4).

Two extra loops (dashed lines) can be applied to optimise the outcome of the guideline in the future. The one loop repeats step 3 which allows additional feed-back for the final scenarios in step 4, the other one reflects the entire methodology and the question of how to improve it further. This will allow a constant adaptation of the guideline according to gained experiences and changing requirements.

TheTest Run

Guidelines determine a certain course of action intending to streamline and optimise processes. The target of the proposed guideline is to prepare coastal cities for sea-level rise and other flooding events. For that, the guideline suggests a specific, integrative way that has the potential to reduce risk and the create new opportunities. A demonstration of this guideline's potential plausibility will happen in the following test run.

Insights created during the test run will check the logical structure of actions and outcomes as well as its content and depth. To test the proposed guideline as effective as possible, the following criteria for the test run's site selection have been considered:

- Already existing hazards like sea-level rise and flooding which are expected to increase in the future even more
- The site's dimension should be not too small (not less than 10 hectares) but still manageable
- Already existing functions on site should be divers to allow urban, green, and blue transformations
- Accessibility for detailed investigations on site should be possible
- Opportunities for future attractive developments (accessibility, security, environmental quality, water views, etc.)

The overlay of these selection criteria led to a specific test site that will be presented in the following first step of the test run.

The test site is located at the City of Gold Coast in Queensland, Australia. Selection parameters included the proximity to a river with flooding challenges and an ocean experiencing a sea-level rise. Moreover, the low-lying and diverse urban character, which combines industrial and residential functions, are providing various design opportunities. The authors would like to clarify that neither the community nor stakeholders selected this specific site. Instead, it was selected by the authors as a test run model.

1. Analysis

The first step of the test run will analyse the selected site to collect as much information about the site as possible. This will happen in three consecutive sub-steps starting with a more general (1.1) to a more specific point of view regarding flood risks (1.2) concluding in a comparison of risks and opportunities (1.3).

The following questions will be thereby examined:

1.1. Where is the test site located, what is the urban context, and how was its historical development? How can the existing functions as well as the built and natural environment be described?

1.2. What is the past, present, and future impact of water due to sea-level rise / storm-water and how can this be simulated?

1.3. Considering the outcome of 1.1. and 1.2., what are the risks and opportunities of the specific areas on the site?

The results will be essential to understand and identify the available elements (urban and natural) on the site, its development throughout the years, the current urban condition, and future hazards and potentials.

© The Author(s), under exclusive license to Springer Nature Singapore Pte Ltd. 2023   1
J. Baumeister and D. Linaraki, *Cities+1m*,
SpringerBriefs in Architectural Design and Technology,
https://doi.org/10.1007/978-981-19-1376-1_1

1. 1 Site Analysis

1.1.1. Site location

N

(Google Earth altitude10km)

N

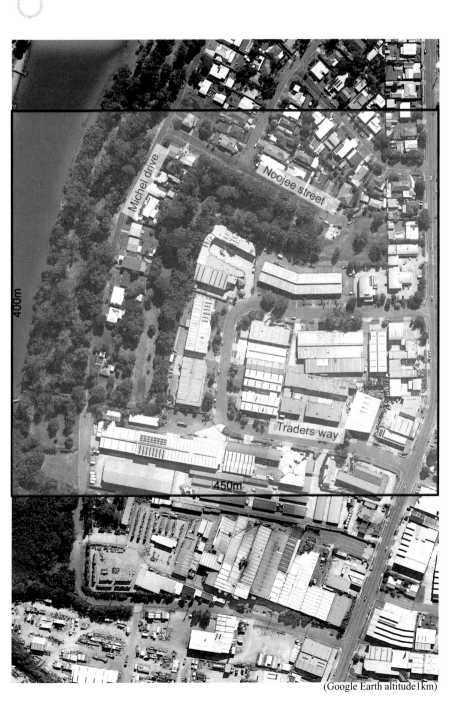

(Google Earth altitude 1km)

1.1.2. Historical development

1930

Old river bed line   Old water passage

1969

N

1974

| New river bed line | New water passage |

Satellite Image reference (QImagery, 2020)   2004

1.1.3. Site visit

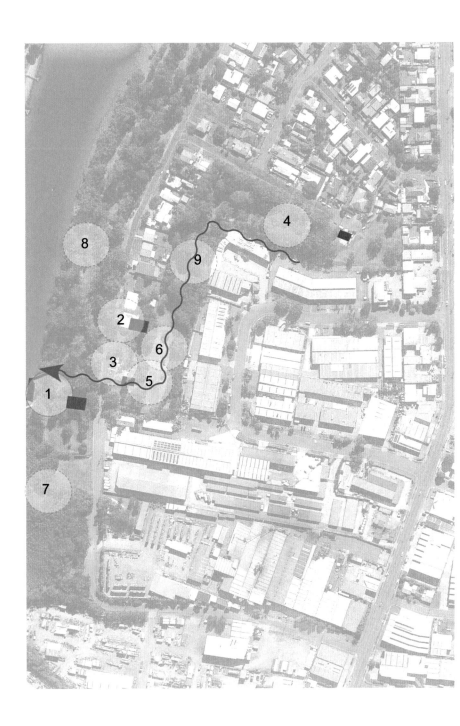

1. Floating platform

5. Dike for stormwater

2. Raised on stilts

6.Muddy flood plain

3. House on wheels

7. Mangroves

4. Basketball

8. Oysters

1.1.4. Buildings, vegetation, elevation

Site plan

Section   100m

N

Flood elevation reference (The City of Gold Coast, 2020)

*HAT (Highest Astronomical Tide) describes the highest
water level that can be expected to occur under average
meteorological conditions

+2.00 elevation
+1.00 elevation
+0.20 elevation
 0.00 HAT*

1.2. Analysis of potential flood risks

1.2.1. Impact of SLR

WATER

A holistic understanding of the urban area will be considered in 2.3. It will include the buildings and urban spaces, production and economy, community, infrastructure and the natural environment. Before, the analysis will focus on the driving element of this guideline which is the impact of water on the site:

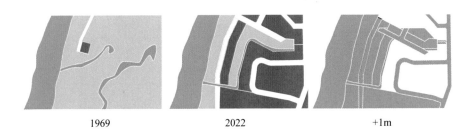

1969   2022   +1m

The historical analysis shows the development on site from a natural water passage to a modified channel to an increasing flooding area.

- The challenges regarding future flooding events and sea level rise are already obvious. There are flooded areas visible and the problems will increase with sea-level rise and incremental growing flooding events especially for buildings and infrastructure which are close to the river. It is also questionable how far the natural environment on site can adapt to potentially fast-rising water levels.

+ The opportunities regarding the future flooding events and sea level rise are also growing with the increasing impact of water: If managed right, proximity to water is creating higher land values, more human health and new business opportunities.

1.2.2. Future flood levels

Current HAT

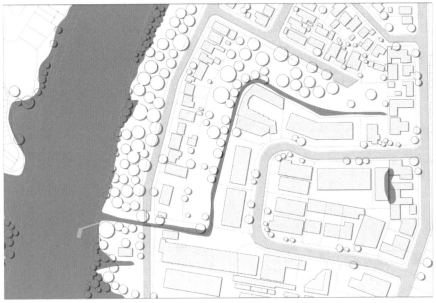

HAT +0.2m SLR

N

HAT +0.5m SLR

HAT +0.8m SLR

1.2.2. Future flood levels

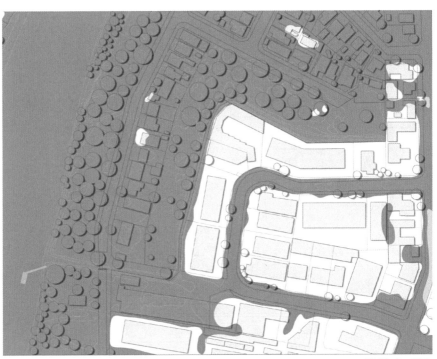

HAT +1m SLR

Flood elevation heights reference
(The City of Gold Coast, 2020)

1.2.3. Risk map

N

High Risk   High to   Medium Risk   Medium to   Adaptable
                Medium Risk   Low Risk   Infrastructure
                                                                    -Raised, Floating,
                                                                    Mangroves

1.3. Analysis of risks and opportunities

1.3.1. Subdivision of site

An understanding of the site's different flooding levels in combination with the current land use allows a subdivision of the area into sub-areas for a more detailed investigation of the site's risks and opportunities.

1. River area   -

2. Vegetation   14850 m²   6. Industrial   20400 m²

3. Vegetation   6400 m²   7. Industrial   21060 m²

4. Residential   11230 m²   8. Residential   12450 m²

5. Vegetation   17705 m²   9. Industrial   25400 m²

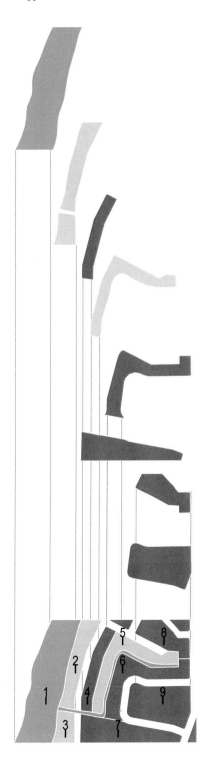

1.3.2. Risks and opportunities*

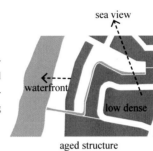

BUILDINGS/SPACE

Residential, commercial, industrial, cultural, and special building typologies in combination with the surrounding urban space

- Flooding issues in the western part
+ Potential for urban development in eastern part due to under-utilisation and old, low dense building structure
+ Underutilised areas
+ Great sea-views from a certain height
+ Intense water connection to the river

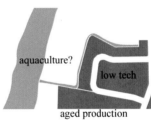

PRODUCTION

Industry, commerce, retail, aquaculture, agriculture, and forestry

+ Aged production facilities enable future production development
+ Proximity to water allows aqua-cultural production centres

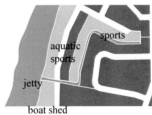

COMMUNITY

Culture, spirituality, education, research, recreation, governance, and politics

- In the eastern part just industrial usage with limited community
+ Western part meeting space for dog owners and aquatic sports including community shed for boats
+ Public jetty
+ Sports field in the central part

* Structure of urban elements is defined in book "Urban Ecolution" (Baumeister and Ottmann, 2014)

INFRASTRUCTURE

Combination of transport, water, sewage, waste, power, and communication

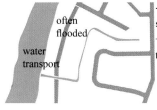

- Western street and related infrastructure endangered by flooding
+ Potential for extended water transport

NATURAL ENVIRONMENT

Amalgamation of sun, air, biomass (plants, animals), minerals, and water

+/- Park with extensive tree coverage along the north-western riverbank which will be endangered by future flooding
+ Area with extensive mangrove forest in south-western part
+ Central green area and artificial river have potential to become a wetland

--- **SUMMARY**

OPPORTUNITIES

Main challenges of the site are aged production facilities in the east and flooded infrastructure in the west

O(A): If required, protective seawalls or dikes can be upgraded with infrastructure, energy, food production

O(B): Great potential for urban upgrading with great sea views and water connection to the river

O(C): Western part provides natural community places with mangroves and wetland opportunities

O(D): Great possibilities for new production facilities including aquaculture and water development

1.3.3. Conclusion of step 1

This first step analysed the current urban and natural elements of the selected site, the flood levels, and the potential risks and opportunities. The analysis helps to understand the past, present and future potentials of the site with the following outcome:

1.1. Following the selection criteria, the site represents a diversity of functions like industry, residential buildings and natural environment. The site visit allowed us to experience the place and to identify already existing flooding areas and floating or raised structures that accommodate current floods. Evaluated building and vegetation heights of the site are thereby based on satellite images and data collected during the site visit conducted.

1.2. The historical analysis of the site reveals its development in a partly muddy area prone to flooding. The initial watercourse was reclaimed and redirected to another location which is a reason for the current flooding challenges. Contour lines were modelled through a combination of satellite images from Google Earth and the Interactive map of the City of Gold Coast (The City of Gold Coast, 2020) which allowed simulations for various flood levels in the future. The consecutive risk map quantified multiple water hazards which require urgent action.

1.3. The subdivision of the test site into manageable parts allowed a detailed description of risks and opportunities. The former includes flooding issues and aged production facilities, the latter potentials for urban upgrading, sea-views and connectivity to water, as well as an integration of green and blue nature.

2. Concepts and Tactics

The achieved understanding of the site and its flooding issues permits the next step which will investigate the sites' potential urban transitions. Following the aim to demonstrate transformations of coastal cities that create more opportunities than threats, the guideline proceeds with investigations regarding the following questions:

2.1. Which different adaptation methods are feasible? And how can they be translated into water-adaptive development concepts which are specific for the test site?

2.2. Which complementary water-features on the urban design scale are imaginable for the test site?

2.3. How can the most appropriate concepts (2.1.) and tactics (2.2.) be combined and visualised?

The results will express theoretically and visually prospective futures for the site, which will be evaluated and improved afterwards.

© The Author(s), under exclusive license to Springer Nature Singapore Pte Ltd. 2023   23
J. Baumeister and D. Linaraki, *Cities+1m*,
SpringerBriefs in Architectural Design and Technology,
https://doi.org/10.1007/978-981-19-1376-1_2

2.1. Concepts

2.1.1. Adaptation methods

Either hard respond as dikes, seawalls, breakwaters, barriers and barrages, or sediment-based as beach, dunes

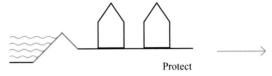

Protect

Raising buildings on stilts and podiums or moving water-sensitive functions of buildings into podiums

Accommodate

Retreat zones, rolling easements, upland buffers are increased, and threatened buildings relocated

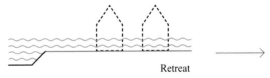

Retreat

Floating structures like buildings or garden, farms are introduced, or land reclamation by land-filling

Advance

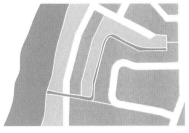

Minimal Change

Refer to 1.3.2. Risks and opportunities O(A)

Protect existing structures

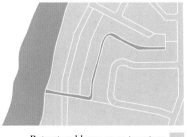

Maximal Yield

Refer to 1.3.2. Risks and opportunities O(B)

Accommodate new structures

Maximal Nature

Refer to 1.3.2. Risks and opportunities O(C)

Retreat and leave areas to nature

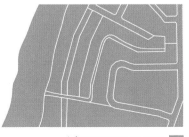

Maximal Water

Refer to 1.3.2. Risks and opportunities O(D)

Advance water on areas

2.1.2. Combinations

Potential Combinations

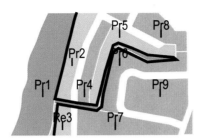

Minimal Change (all areas kept and protection structure added)

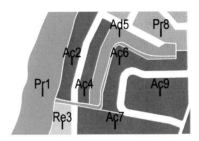

Maximal Yield (most areas redeveloped, inner harbour included)

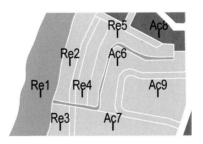

Maximal Nature (flooded areas transferred into natural wetland)

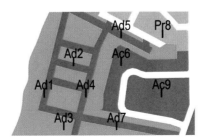

Maximal Water (additional water areas for extra functions added)

2.2. Tactics

2.2.1. Tactics' logos

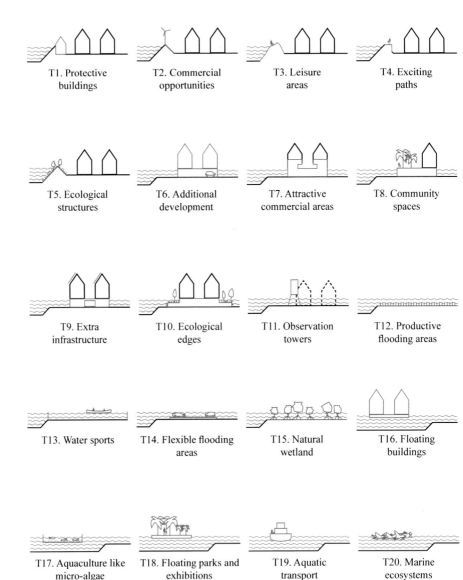

| T1. Protective buildings | T2. Commercial opportunities | T3. Leisure areas | T4. Exciting paths |

| T5. Ecological structures | T6. Additional development | T7. Attractive commercial areas | T8. Community spaces |

| T9. Extra infrastructure | T10. Ecological edges | T11. Observation towers | T12. Productive flooding areas |

| T13. Water sports | T14. Flexible flooding areas | T15. Natural wetland | T16. Floating buildings |

| T17. Aquaculture like micro-algae | T18. Floating parks and exhibitions | T19. Aquatic transport | T20. Marine ecosystems |

2.2.2. Tactic's explanations

The water-adaptive concepts determined before will support the different planning approaches. In addition to that, more detailed features on the urban design scale will support with potential tactics (Baumeister, 2020) like:

T1. Add buildings to protective structures which expand the urban space
T2. Use the surface of protective measures for commercial opportunities
T3. Create landscapes on protective structures which offer leisure areas
T4. Generate attractive trails to walk and cycle on the top of dykes
T5. Contribute with plants to the natural diversity of flora and fauna

T6. Produce additional attractive interior and exterior space on podiums
T7. Use additional with functions for industry, commerce and retail
T8. Create public spaces on podiums as opportunities for communities
T9. Keep infrastructure centralized or change it into decentralised
T10. Upgrade the ecology between land and water with amphibious zones

T11. Equip floodplains with observation towers or convenient buildings
T12. Grow aquaculture products in flood areas such as rice or micro-algae
T13. Define temporary leisure and community opportunities in floodplains
T14. Use e.g. parking lot as zones which can be flooded in an emergency
T15. Support amphibious ecosystems and indigenous vegetation

T16. Extend urban uses and spaces with attractive floating buildings
T17. Benefit from proximity to water for commercial production facilities
T18. Expand recreation areas and cultural institutions onto the water.
T19. Develop opportunities for floating transport and infrastructure
T20. Contribute to a floating amphibious and aquatic ecological diversity

2.3. Concept + Tactics = Scenarios

2.3.1. Opportunity A

Opportunity A: Minimal Change

To succeed in minimal change, protection elements are used in locations that are identified as high risk based on the risk map information. Parts of the site that is currently composed of mangrove forests will retreat so that the mangrove forest can expand.

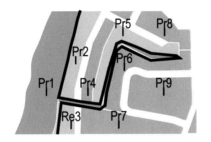

Concepts: Combination of "Protect" for the areas 1, 2, 4, 5, 6, 7, 8, 9 with "Retreat" for area 3

Tactics: T1. Protective buildings or infrastructures, T3. Leisure areas, T4. Exciting paths, T5. Ecological structures, T15. Natural wetland, T19. Aquatic transport

Created Opportunities: O(A). (see risks and opportunities) Protective sea-walls and dikes in combination with O(C). Mangroves and wetland opportunities

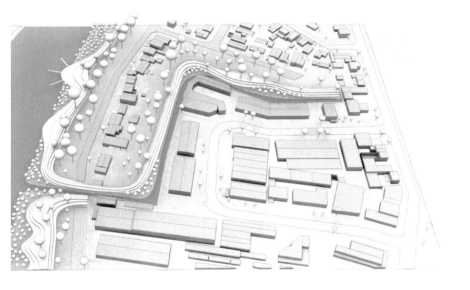

2.3.2. Opportunity B

Opportunity B: Maximal Yield
The existing conditions will have to accommodate to the +1m water level including
areas 2, 4, 6, 7 and 9 which are proposed to be equipped with buildings construct-
ed on podiums. Area 5 is designed to collect water and area 3 retains the existing
mangroves.

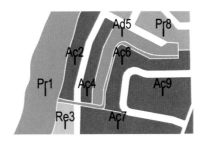

Concepts: Combination of "Accommodate" for areas 2,4,6,7,9 with "Protect" for
1,8, "Retreat" for 3 and Advance for the area 5

Tactics: T6. Extra development, T7. Attractive commercial areas, T8. Community
spaces, T9. Extra infrastructure, T10. Ecological edges, T15. Natural wetland, T19.
Aquatic transport

Created Opportunities: O(B). Exceptional potential for attractive urban upgrading
with great sea views and water connection to the river with O(C). Western part provides natural community places with mangroves and O(D). Great possibilities for a
water development like a harbour

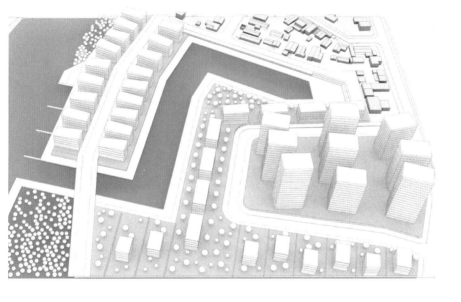

2.3.3. Opportunity C

Opportunity C: Maximal Nature
To succeed with the maximal natural environment, all areas are proposed to retreat and eventually become a wetland. However, areas 6, 7 and 9 are designed to accommodate residential towers raised on stilts so that the natural environment can expand at the ground level.

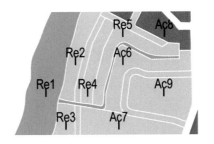

Concepts: Combination of "Retreat" for areas 1,2,3,4,5 in combination with "Accommodate" for the areas 6,7,8,9

Tactics: T4. Exciting paths, T6. Extra development, 8. Community spaces, 12. Productive flooding areas, 13. Water sports, 14. Flexible flooding areas, 15. Natural wetland

Created Opportunities: O(B). Great potential for urban upgrading with great sea views and water connection to the river and O(C). The western part provides natural community places with mangroves and wetland opportunities

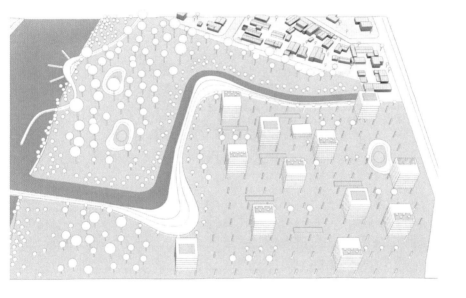

2.3.4. Opportunity D

Opportunity D: Maximal Water
To succeed with maximal water, this proposal allows the water to extend towards the inland. The existing areas of 1,2,3,4,5,7 are redesigned to a floating development that supports the natural flow of water. New functions for areas 6 and 9 are accommodated by raising the land to a minimum of 1m height.

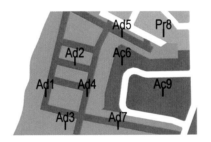

Concepts: Combination of "Advance" for areas 1,2,3,4,5,7, "Accommodate" for 6,9 and "Protect" for area 8

Tactics: T1. Protective buildings or infrastructures, T6. Extra development, T8. Community spaces, T16. Floating buildings, T17. Aquaculture like micro-algae, T18. Floating parks and exhibitions, T19. Aquatic transport, T20. Marine ecosystems

Created Opportunities: O(B). Great potential for urban upgrading with great sea views and water connection to the river, O(C). The western part provides natural community places with mangroves and wetland opportunities, and O(D). Great possibilities for new production facilities

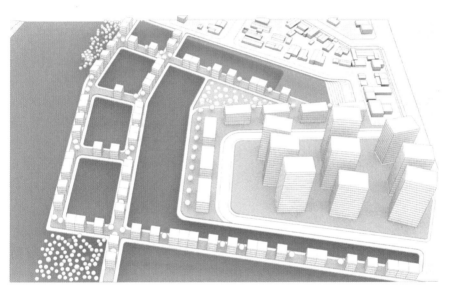

2.3.5. Conclusion of step 2

Step 2 was taking the outcomes of step 1 and introduced adaptation methods which were developed further to concepts and tactics. Each of the sub-steps 2.1., 2.2. and 2.3. were introducing thereby detailed approaches:

The adaptation methods were transferred into different development scenarios for the site which were diversified on the site's different areas. Finally, individual water-related tactics were identified and introduced on the site which completed the development process from strategic planning to urban planning to an urban design scale.

The selected combinations of concepts and tactics are only one possible choice. It seems to be the most beneficial one for each concept but this has to be tested with stakeholders and the community. The resulting visualisations and explanations of step 2 showcase the four chosen concepts which will be developed further to scenarios after having received the feed-back of the following step 3.

3. Feed-back

Until now, the guideline created four visualised concepts and considered 20 tactics which resulted in four scenarios. The target of this step 3 is to test this theoretical outcome under "real conditions" by asking stakeholders for a feed-back. The composition of the stakeholders as well as the feed-back process should follow each project's individual requirement to keep it as efficient as possible.

Therefore the following feed-back of the guideline's test-run is representing just one possible way. The corona situation permitted this test-run as an online format collecting feed-back of city council members.

The following pages will represent some answers to the key questions:

Are the four suggested concepts equally attractive for future considerations?

Are the twenty suggested tactics equally attractive for future considerations? Or which tactics can be eliminated in the future process?

Evaluating the four different visualised concepts, what advantages and disadvantages are coming into your mind? Do you have any additional comments?

For purposes of clarity, each question (3.1.) will respond directly to the feed-back (3.2.).

© The Author(s), under exclusive license to Springer Nature Singapore Pte Ltd. 2023   41
J. Baumeister and D. Linaraki, *Cities+1m*,
SpringerBriefs in Architectural Design and Technology,
https://doi.org/10.1007/978-981-19-1376-1_3

3.1. Survey

3.1.1. Survey of Adaptation methods

Protect

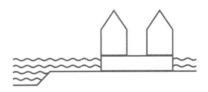

Accommodate

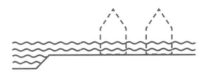

Retreat

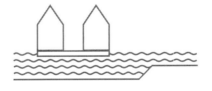

Advance

3.1.2. Survey of Tactics

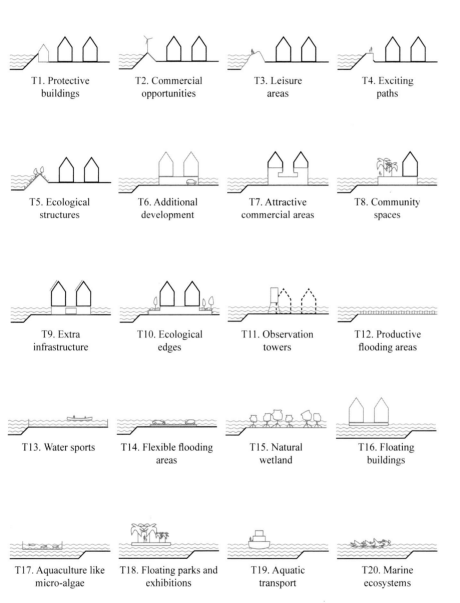

T1. Protective buildings

T2. Commercial opportunities

T3. Leisure areas

T4. Exciting paths

T5. Ecological structures

T6. Additional development

T7. Attractive commercial areas

T8. Community spaces

T9. Extra infrastructure

T10. Ecological edges

T11. Observation towers

T12. Productive flooding areas

T13. Water sports

T14. Flexible flooding areas

T15. Natural wetland

T16. Floating buildings

T17. Aquaculture like micro-algae

T18. Floating parks and exhibitions

T19. Aquatic transport

T20. Marine ecosystems

3.1.3. Survey of Scenarios

Minimal Change   Maximal Yield

Advantages   Advantages

+ Improved amenity and social out-   + Cost of adaptation transferred to de-
comes   velopers
+ Lowest risk of unintended off-site   + Maximum yield
impacts from creating new waterbodies   + Economic benefits from the develop-
+ Minimal capital expenditure required   ment
                                                  + Waterfront amenity benefits with
                                                  high water quality (tidal)

Disadvantages   Disadvantages

- Loss of land for employment and ac-   - Exposing a greater amount of devel-
commodation (displacement of local   opment to natural hazards
jobs and housing)   - Environmental systems impacted
- Access issues (flood inundation)   - Potential for unintended off-site im-
- Infrastructure problems (existing sewer   pacts to the environment (change to
systems may overflow into waterways)   tidal watercourses)
                                                  - Difficulty to assemble properties
                                                  - Very high capital expenditure re-
                                                  quired
                                                  - Regulatory process issues with cre-
                                                  ating new waterbodies

3.2. Results

3.2.1. Results of Adaptation methods (example)

Advantages	Disadvantages
+ Consistent with current approach	- Can be overtopped and cause damage
+ Allows for natural processes	- Creates risk of isolation
+ Removes exposed assets from zone of impact	- Loss of developable area
+ Robust and flexible	- Administrative barriers

3.2.2. Results of Tactics (example)

+ / −	**+** / −	**+** / −	**+** / −
T1	T2	T3	T4

+ / −	**+** / −	**+** / −	**+** / −
T5	T6	T7	T8

+ / −	**+** / −	**+** / −	**+** / −
T9	T10	T11	T12

+ / −	**+** / −	**+** / −	**+** / −
T13	T14	T15	T16

+ / −	**+** / −	**+** / −	**+** / −
T17	T18	T19	T20

3.2.3. Results of Scenarios

Maximal Nature   Maximal Water

Advantages   Advantages

+ Development accommodates space
for water movement and flow
+ Benefits of natural environment
and waterfront amenity (subject to
water quality being achieved)
+ Minimal capital expenditure required and reduced need for site as-
sembly

+ Benefits of waterfront amenity (subject to water quality being
achieved)
+ Economic benefits from development and high yield

Disadvantages   Disadvantages

- Potential access issues (flood inundation)
- Loss of land for employment and
accommodation (displacement of
local jobs and housing)

- Lots of additional infrastructure to
maintain
- Environmental systems impacted
- Unintended off-site impacts on the
environment
- Difficulty to assemble properties
- The highest level of capital expend-
iture required
- Regulatory process issues with cre-
ating new waterbodies

3.2.4. Conclusion of step 3

The goal of the feed-back round was to test the guideline's steps 1 and 2 regarding logic, comprehensibility, and level of detail provided with the following outcome:

The first part of the assessment finds advantages (and disadvantages) of all 4 concepts. Therefore there is no reason to exclude one concept in the final scenarios.

All proposed tactics were evaluated very positively in the second part of the feed-back, except for three tactics which are related to the retreat concept. These three tactics won't be considered in the final scenarios.

A detailed comparison between the advantages and disadvantages of the assessment's third part confirms the outcome of part one. The visualisation of concepts and tactics works very well in terms of comparability and depth of details and the assessed stakeholders were able to participate by putting own knowledge and opinions into the process.

Therefore the feed-back collected knowledge which helped to approve the set-up of the guidelines which allows proceeding to the final step.

4. Final Scenarios

This final step 4 will develop the four concepts further into four scenarios. It will detail the visualised concepts of step 2 and will take the preferences of step's 3 feedback into account.

The positive evaluation of all concepts suggests not to exclude one which will create a diverse variety of scenarios. The tactics can be shown more specified due to the larger scale of the drawings.

The point of view of the perspectives will be identical to be able to crosscompare the different scenarios. This should help to demonstrate the flexibility of the urban development solutions and the potentials to mix in the future different scenarios with different areas.

Therefore following visualisations are not final proposals but atmospheric stimulations which want to showcase the variety of solutions for Urban Development Solutions for Sea-level Rise for further discussion between stakeholders and the community.

© The Author(s), under exclusive license to Springer Nature Singapore Pte Ltd. 2023   51
J. Baumeister and D. Linaraki, *Cities+1m*,
SpringerBriefs in Architectural Design and Technology,
https://doi.org/10.1007/978-981-19-1376-1_4

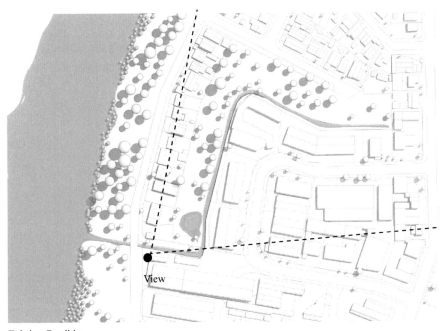

View

Existing Condition

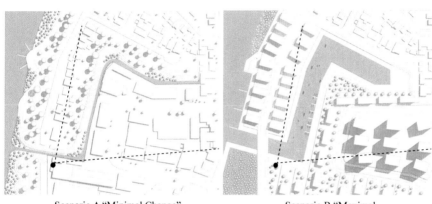

Scenario A "Minimal Change"   Scenario B "Maximal

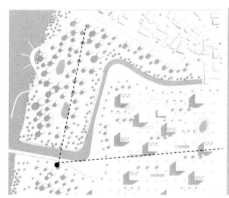
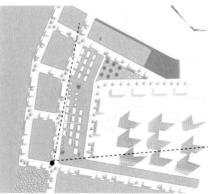

Scenario C "Maximal Nature"   Scenario D "Maximal Water"

4.1. Existing Condition

4.2. Scenario A "Minimal Change"

4.3. Scenario B "Maximal Yield"

4.4. Scenario C "Maximal Nature"

4.5. Scenario D "Maximal Water"

5. Research Outcome

Following results derived based on the conclusions of steps 1-3 and the outcome of step four:

a. The guideline creates multiple answers to the questions "Which concept should be applied to transform coastal cities to +1m and who will take the decisions?": Four final scenarios have been described and visualised. Each scenario demonstrates thereby different opportunities for the required transformation of coastal cities due to sea level rise. The scenarios are comparable in terms of advantages and disadvantages for stakeholders' discussions and decisions which enables a diverse involvement of various decision-makers.

b. The guideline considers state-of-the-art processes and is therefore forward-looking: It is flexible in terms of decision making and adaptable to future demands. It could be applied in different planning contexts, promotes sustainable planning guidance, considering specific site conditions and water-hazards, and stimulating the connection of government and community into research and education.

c. The guideline broadens the perspective and turns challenges into opportunities: This methodology promotes an integration of various disciplines which allows to work with nature and not against it. This shifts the perspective from an approach that wants to solve the problem of sea-level rise technically, to urban transformation scenarios for coastal cities. This opens up space for grey, green, or blue solutions for an adaptable future urban development.

J. Baumeister and D. Linaraki, *Cities+1m*,
SpringerBriefs in Architectural Design and Technology,
https://doi.org/10.1007/978-981-19-1376-1_5

The research outcome provides opportunities for a continuation of the guideline towards implementation (refer to 5.) as a next step.

4   Final Scenarios

Step 4 (finalised): The manual and test run ends here. Different development scenarios (minimal change, maximal yield, maximal nature, maximal water) have been developed in parallel together with a rough understanding of the related pros and cons.

5   Feasibility study for each Scenario

Step 5: Each developed scenario will be examined further. This should include an integrated planning approach with workshops on all governmental levels and disciplines under the involvement of representatives from community, industry and research.

6   Decision for one, or a combined Scenario

Step 6: Stakeholder workshops that cross-compare the feasibility of scenarios. An assessment of risks and opportunities will end in the prioritisation of one scenario or in the development of a combination of the scenarios on the site's separate areas.

7   Formal Planning, Design, Implementation process

Step 7: Normal planning approach supervised by a committee which controls the quality of the scenario's implementation.

References

Baumeister, J. (2020). Re-Building Coastal Cities: Urban Systematics for 20 Sea-Level Rise Tactics. In Joerg Baumeister, Edoardo Bertone, & P. Burton (Eds.), Sea Cities. Urban Tactics for Sea-Level Rise: Springer.

Baumeister, J., & Ottmann, D. A. (2014). Urban Ecolution: A pocket generator to explore future solutions for healthy and ecologically integrated cities: University of Western Australia.

Cambridge. (2020). Retrieved from https://dictionary.cambridge.org/

Chrysoulakis, N., Castro, E. A. d., Moors, E. J., & ProQuest, E. (2015). Understanding urban metabolism: a tool for urban planning. Abingdon, Oxon : New York, NY: Routledge.

Jones, N. (2017). How the World Passed a Carbon Threshold and Why It Matters. In Yale Environment: Yale School of the Environment.

Oppenheimer, M., B.C. Glavovic , J. Hinkel, R. van de Wal, A.K. Magnan, A. Abd-Elgawad, R. Cai, M. Cifuentes-Jara, R.M. DeConto, T. Ghosh, J. Hay, F. Isla, B. Marzeion, B. Meyssignac, and Z. Sebesvari. (2019). Sea Level Rise and Implications for Low-Lying Islands, Coasts and Communities. In: IPCC Special Report on the Ocean and Cryosphere in a Changing Climate.

Queensland Goverment, (2020). QImagery. Retrieved from https://qimagery.information.qld.gov.au/

Samset, B. H., Fuglestvedt, J. S., & Lund, M. T. (2020). Delayed emergence of a global temperature response after emission mitigation. Nature communications, 11(1), 1-10. doi:10.1038/s41467-020-17001-1

The City of Gold Coast (2020). City Plan interactive mapping - Version 8. Retrieved from http://cityplanmaps.goldcoast.qld.gov.au/CityPlan/

Notes

Cities+1m is the result of an interdisciplinary research team led by Prof. Dr.-Ing. Joerg Baumeister and PhD Researcher Despina Linaraki.

We would like to acknowledge, with gratitude, the contributions of various city council members, the broader community of SeaCities Lab and Griffith University.

Photos (except of the images used for the historical development), diagrams, and plans have been produced by Joerg Baumeister and Despina Linaraki. Copyright and all images remain property of SeaCities and may not be reproduced without permission.

Printed in the United States
by Baker & Taylor Publisher Services